EXPLORING ENGINEERING CAREERS

Exploring CIVIL ENGINEERING Careers

Carla Mooney

ReferencePoint Press

San Diego, CA

ReferencePoint Press®

Printed in the United States

For more information, contact:
ReferencePoint Press, Inc.
PO Box 27779
San Diego, CA 92198
www.ReferencePointPress.com

LIBRARY OF CONGRESS CATALOGING-IN-PUBLICATION DATA

Name: Mooney, Carla, 1970- author
Title: Exploring Civil Engineering Careers / by Carla Mooney.
Description: San Diego, CA : ReferencePoint Press, 2027. | Series: Exploring Engineering Careers | Includes bibliographical references and index.
Identifiers: LCCN 2025049478 (print) | LCCN 2025049479 (ebook) | ISBN 9781678212865 library binding | ISBN 9781678212872 ebook
Subjects: LCSH: Civil engineering--Vocational guidance--Juvenile literature | Civil engineers--Juvenile literature | CYAC: Civil engineers | Civil engineering--Vocational guidance | LCGFT: Informational works
Classification: LCC TA157 .M66 2026 (print) | LCC TA157 (ebook)
LC record available at https://lccn.loc.gov/2025049478
LC ebook record available at https://lccn.loc.gov/2025049479

CONTENTS

Civil Engineer at a Glance 4

Introduction 5
Rebuilding the Fern Hollow Bridge

Chapter One 9
What Does a Civil Engineer Do?

Chapter Two 17
How Do You Become a Civil Engineer?

Chapter Three 25
What Skills and Personal Qualities Matter Most—and Why?

Chapter Four 32
What Is It like to Work as a Civil Engineer?

Chapter Five 39
Advancement and Other Job Opportunities

Chapter Six 45
What Does the Future Hold for Civil Engineers?

Source Notes 53
Interview with a Civil Engineer 56
Find Out More 59
Index 60
Picture Credits 63
About the Author 64

Civil Engineer at a Glance

Working Conditions

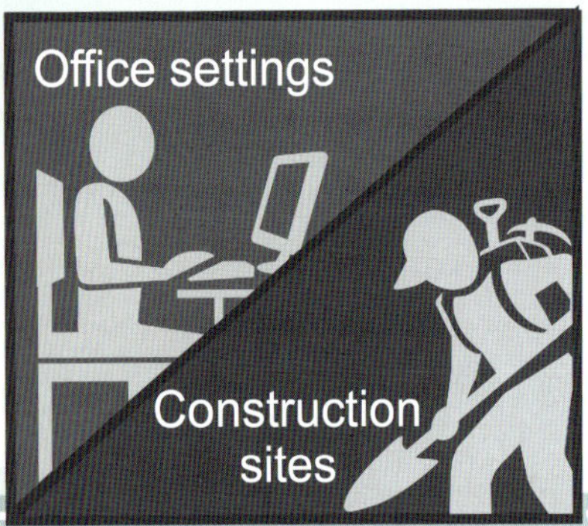

Certification and Licensing

Professional Engineer (PE) license required to provide services directly to the public

Personal Qualities

- Good communication
- Organized
- Decision-making
- Problem-solving
- Interpersonal
- Math skills

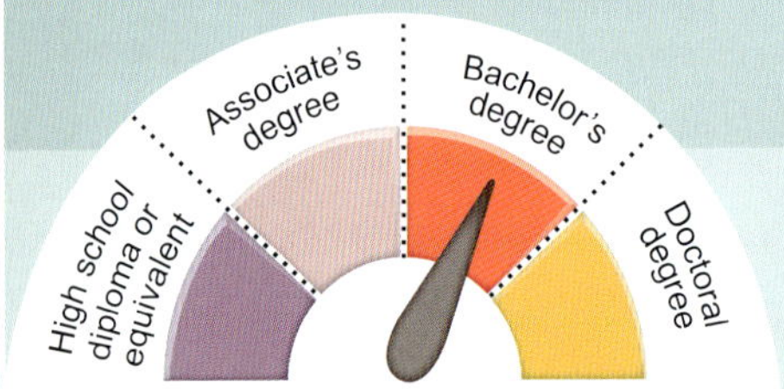

Minimum Educational Requirements

Bachelor's degree

Pay

$99,590 median pay in 2024

Future Job Outlook

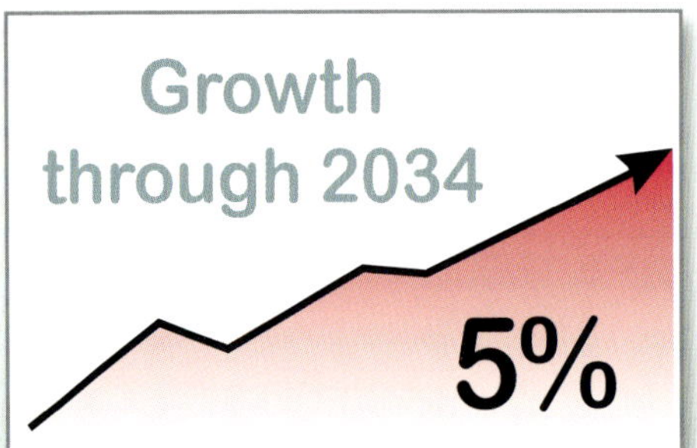

Number of jobs in 2024

368,900

Source: US Bureau of Labor Statistics, "Civil Engineers," *Occupational Outlook Handbook*, August 28, 2025. www.bls.gov.

Rebuilding the Fern Hollow Bridge

On January 28, 2022, some residents of Pittsburgh, Pennsylvania, awoke to a loud noise shortly before 7:00 am. "There was a boom, then a monster sound. It was so loud, and it didn't stop,"[1] says Melissa Bakth, who lives near the city's Frick Park and the Fern Hollow Creek. What Bakth heard was the sound of the Fern Hollow Bridge collapsing into a ravine. The 447-foot (136 m) bridge, opened in 1973, was a busy four-lane roadway that spanned the Fern Hollow Creek.

Immediately after the collapse, safety was the top priority. Six vehicles were on the bridge when it collapsed. First responders rescued those trapped, with several people injured and no fatalities. They secured and inspected the site to ensure there was no immediate danger to nearby residents and commuters. A National Transportation Safety Board investigation later determined that the bridge collapsed due to structural failure resulting from lapses in regular maintenance.

Building a New Bridge

Within five days of the collapse, planning for a replacement bridge began. The Fern Hollow Bridge was a vital roadway in the community, and state and local officials deemed the replacement project to be a top priority. Civil engineers with bridge expertise were invited to join the project team and design the replacement bridge.

Senior project manager Jason Fuller led engineering firm HDR's work on the new bridge. The goal was to design and build a new bridge as quickly as possible while maintaining safety. To do so, Fuller's design team worked closely with the construction team to be as efficient as possible. "Infrastructure design projects usually involve linear processes. In this case, given the urgency, the design team worked with PennDOT [Pennsylvania Dept. of Transportation] to determine what could be done in parallel, where data already available could be used, and when partial submissions could be submitted for review, rather than complete submissions,"[2] says Fuller.

Fuller and HDR worked on the bridge design from the top down, which is typical for bridge design, while also working from the bottom up. They used knowledge of local geology and recent geotechnical data from a nearby bridge replacement project in their design. "Designing the bottom of the bridge and its foundation elements was important, given the speed of the project, as they were among the first items that needed to be constructed,"[3] says Fuller.

To expedite construction, the design team worked closely with the bridge contractor to create a design that used materials and equipment that were on hand or readily available. They prioritized completing designs of components that would be needed first, such as beams and bearings, which had long order times, and foundations, because they would be the first items built. This just-in-time design approach kept the project moving forward and the construction timeline on track. "Overall, the design-build process facilitated an incredible awareness of the effects of design decisions on construction and vice versa, which was only heightened by the high profile of the project,"[4] says Fuller.

Fuller and his design team also gathered community input for the new bridge design. They incorporated community requests for improved bicycle and pedestrian lanes into the design while maintaining the required number of traffic lanes on the bridge.

On December 21, 2022, state and local officials gathered to cut the ribbon on the new bridge, just 327 days after the collapse. The next day, the bridge opened to a single lane of traffic. Work continued on some components, including lighting and a pedestrian crossing. In July 2023 the bridge fully reopened. Bill James, a Pittsburgh resident and frequent user of the bridge, expressed his gratitude to those who worked to replace the collapsed bridge so quickly. "I am just appreciative of all the people who put the hard effort into designing the bridge and also building it under such a short time frame. I think it is a great testament to those both in the design and laboring community. . . . They were able to pull together and make a great impact for our community,"[5] he says.

> **"I am just appreciative of all the people who put the hard effort into designing the bridge and also building it under such a short time frame. I think it is a great testament to those both in the design and laboring community. . . . They were able to pull together and make a great impact for our community."[5]**
>
> **—Bill James, Pittsburgh resident**

Engineering Improves Lives

The replacement of the Fern Hollow Bridge shows how engineering can solve real-world problems and improve human lives. Engineering is present in nearly every aspect of modern life. Smartphones, medical devices, cars, renewable energy systems, clean water access, and sustainable buildings are all products of engineering. Without engineers, many modern conveniences and essential systems that people rely on every day would not exist.

Engineering is a broad field that includes many disciplines, from developing new technologies to designing skyscrapers. Civil engineering is a branch of engineering that focuses on the design, construction, and maintenance of infrastructure such as roads, bridges, dams, buildings, airports, and water systems. Civil engineering provides the foundation for other types of engineering. For example, civil engineers design the buildings and power grids that electrical engineers rely on to support power systems. Engineers use problem-solving skills, innovation, and technical skills to create practical solutions that make life safer, easier, and more efficient.

Civil engineering plays a critical role in shaping and maintaining the world's built environment. The structures civil engineers create—including roads, water systems, bridges, and buildings—support how people live, work, and travel every day. Now and in the future, civil engineers are at the forefront of designing stronger, safer, and more sustainable buildings and infrastructure for people worldwide.

What Does a Civil Engineer Do?

Throughout history, humans have built incredible structures, from the world's tallest building, the Burj Khalifa in Dubai, to the Hangzhou Bay Bridge in China. Civil engineering is one of the oldest and most important branches of engineering. It focuses on designing, building, and maintaining the structures and systems that support modern life. Civil engineers work on projects such as roads, bridges, tunnels, buildings, airports, dams, and water supply systems. Their work helps make communities safer, stronger, and more efficient.

Building the World

Civil engineers play an essential role in designing, building, and maintaining the structures and systems that support daily life. They use math, science, and technology to solve real-world problems. For example, they figure out how to make bridges strong enough to carry heavy loads, prevent flooding in cities, and build earthquake-resistant buildings. They also consider the environment, safety, and cost when planning their projects. Their work involves many different responsibilities throughout each stage of a project.

Civil engineers are often involved from the earliest planning stages to the final inspection of a project. In the planning phase, civil engineers begin by studying plans, maps, surveys, and other data related to the project's design. They evaluate site conditions, environmental factors, and government regulations to make sure a project is feasible and safe. They also analyze soil and construction materials to determine whether they are strong enough for foundations and support structures. "Every project has a structure to support 'something,' and that 'something' varies quite a bit per project," says Mark McKinney, a civil engineer with the Ross Group, a construction company in Oklahoma. "We do anything from oil and gas, industrial, and commercial buildings, to concrete reaction masses for vibration testing, parking garages, and historical renovations." Even though every project is different, McKinney says there is a general process to be followed. "The first step is to gather all the information on what I'm designing. Next, I consider all the loads and select a structural system. The challenge is to then figure out how to create a set of drawings that will describe the project accurately to someone who knows nothing about it!"[6]

> **"Before a project can be built, it has to exist on paper—our drawings. Before it exists on paper, it only exists in my mind. All of this work starts with my ideas, which eventually become something real in the world that people see and use—and I like seeing my ideas take shape and come to life."[7]**
>
> **—Mark McKinney, civil engineer**

In the design phase, civil engineers use computer-aided design software to create detailed project plans. "Before a project can be built, it has to exist on paper—our drawings. Before it exists on paper, it only exists in my mind. All of this work starts with my ideas, which eventually become something real in the world that people see and use—and I like seeing my ideas take shape and come to life,"[7] says McKinney.

During the planning phase, civil engineers also research building codes, traffic patterns, and material availability to estimate timelines and costs. They prepare cost estimates for materials, equipment, and labor to determine whether a project is affordable. They apply for and obtain required permits before construction can start.

During construction, civil engineers oversee surveying to set up building locations, site layouts, and elevations. Throughout construction, they make sure the project meets design specifications and safety standards. They handle tasks such as inspections and billing and help solve problems that occur on-site. "Sometimes one of our constructors comes to me with a problem and their project is 'stuck,' meaning progress has slowed or stopped," says McKinney. "When I spend some time to solve the issue and get their project going again, they are appreciative. I discover something new on almost every project, so I never stop learning, which is something I love about the job and is important to me,"[8] says McKinney.

Types of Projects

Civil engineers frequently work with architects, construction managers, and urban planners on a variety of projects. They design and oversee the construction of buildings and other structures, including skyscrapers, homes, office buildings, hotels, theaters, and many other types of structures. Engineers work closely with construction firms and developers to ensure that design requirements and safety regulations are met.

Civil engineers also work on environmental projects that address environmental concerns and provide solutions. Examples

The World's Tallest Building

The Burj Khalifa, located in Dubai, United Arab Emirates, is the tallest building in the world, standing at 2,717 feet (828 m). Completed in 2010, the Burj Khalifa is considered a civil engineering marvel because of the innovative techniques used to make it strong, stable, and safe at such an extreme height. Its structure uses a special Y-shaped, buttressed core design that helps support the building's enormous weight and resist strong winds. It has a hexagon-shaped central concrete core and three wings that form its Y-shape and act as buttresses. A buttress is a projecting support, usually made of stone or brick, built against a wall to give it strength and stability. The foundation of the tower consists of massive concrete piles drilled deep into the ground and reinforced with steel to ensure the structure's stability in sandy soil. Advanced materials, including high-performance concrete and steel, help reduce weight and increase strength. Engineers used approximately 11.7 million cubic feet (331,307 cu. m) of concrete and 42,990 tons (39,000 metric tons) of reinforced steel to build the Burj Khalifa.

include creating flood barriers to protect communities and vulnerable habitats or establishing pollution control systems to protect human health. Some projects involve water and wastewater systems, such as dams, sewage treatment plants, and water supply networks. These projects help protect public safety and improve the quality of life in communities. Other times, civil engineers work on infrastructure projects. They are involved in the design and construction of roads, bridges, railways, tunnels, and airports. For example, civil engineers from Kleinfelder, an engineering firm, helped design and construct the New Terminal 1 at San Diego International Airport. Kleinfelder engineers led project teams that focused on bridge, retaining wall, and foundation design. During construction, engineers designed staged construction drawings for a new three-lane roadway to streamline traffic flow around the terminal.

The first phase of the terminal opened in September 2025 with nineteen gates, expanded parking, and roadways designed for efficient access to the airport. A second phase is planned with eleven additional gates. When complete, the new terminal will provide an improved travel experience for more than 11 million passengers each year. "Having lived and worked in San Diego for over 30 years, this is both a personal and professional milestone," says Jim Frost, a landside civil engineering lead on the project. "We're proud to have contributed to this local project that enhances connectivity and efficiency. It's exciting to see our work come to life in a space that will serve millions of travelers, including our family and friends, for many years."[9]

Specialization in Civil Engineering

Some civil engineers are generalists and work on many different types of projects. Others specialize in specific areas of civil engineering, much as some surgeons specialize in specific body systems, such as orthopedics or cardiology. Civil engineering specialties include construction, structural, environmental, transportation, or water resources engineering.

Civil engineers design and oversee the construction of buildings and other structures. They often work with architects, construction managers, and urban planners to ensure that design requirements and safety regulations are met.

As the name suggests, construction engineers specialize in construction projects. They often work as project managers, ensuring that each project is completed according to plans and specifications. They oversee the project from start to finish, ensuring that it is completed on time and on budget and that all safety standards are met.

Structural engineers specialize in designing and analyzing the load-bearing capacity of buildings, bridges, and other structures. Load-bearing capacity is the maximum weight a structure can support without failing or collapsing. Structural engineers also design structures that can withstand nature's forces, such as strong winds, heavy snow, and even earthquakes.

Designing Earthquake-Resistant Structures

Some civil engineers specialize in designing structures that can withstand an earthquake. They use special materials, designs, and technology to help structures withstand shaking and movement. One key method is base isolation, which places flexible bearings or pads between a building and its foundation. This allows the ground to move without transferring all the motion to the structure above. Energy dissipation devices, also known as seismic dampers, spread out the energy produced during an earthquake, reducing the force exerted on the structure. The use of reinforced concrete and steel frames that can bend slightly without breaking reduces the chance of collapse. Cross-bracing and shear walls add extra support to structures. Building on flexible foundations, like a reinforced concrete slab on a sand cushion, and installing adequate draining systems also help maintain stability during an earthquake. Often, civil engineers will test their designs with computer models to see how the structure will respond in different earthquake scenarios.

A structure is only as stable as the ground on which it is built. Geotechnical engineers assess a site's soil and rock to determine whether the site will be safe and stable for a construction project. These engineers perform site investigations and collect soil, rock, and water samples. They perform lab and field tests of soil and rock to determine their physical properties, such as strength, compressibility, and permeability. They identify potential risks—such as landslides, soil settlement, and slope instability—and develop plans to mitigate these risks. Geotechnical engineers also design foundations, retaining walls, and other structures that meet the ground. Their work is essential to building stable, secure structures.

Melinda Eason is a geotechnical engineer with the US Army Corps of Engineers and has worked on a wide range of projects with other civil engineers. She says:

> Geotechnical engineers work with nearly every other civil engineering specialty. Whether a roadway engineer needs to understand pavement conditions, a structural engineer requires information about bearing capacity, or an oil and

gas engineer needs to know information on the water table and soil types, we're involved in nearly every aspect of construction. We're the go-to experts for soil-related issues and work alongside diverse engineers to help build our world.[10]

Transportation engineers focus on the planning, design, operation, and maintenance of systems that move people and goods. They study traffic patterns and forecast future needs. These engineers design and develop roads, highways, railways, airports, and public transit systems. They plan systems and develop solutions to improve traffic flow, manage congestion, and maintain high safety standards. Kelley Kircher has worked in transportation engineering for more than twenty years. "I help get people where they need and want to go, more safely, more efficiently, and using different modes of transportation," she says. "At its heart, that is what Transportation Engineering is all about: math, science, economics, analytics, empathy, vision, writing, and listening. As Sir Isaac Newton said, 'We build too many walls and not enough bridges.'"[11]

"There are as many different types of engineer as you can imagine, but they're all solving people's problems. From ensuring the basics, like plenty of clean water in our taps to designing and building the places we live, work and play in, all over the world."[12]

—Rob Ridley, civil engineer

Some civil engineers focus on environmental engineering or water resources. Those who focus on environmental engineering often work on sustainable projects, such as designing green buildings, water purification systems, wastewater treatment systems, renewable energy projects, and recycling systems. They ensure projects comply with environmental regulations. Civil engineers who specialize in water resources design dams, canals, and stormwater management solutions.

Civil engineering is a broad career that involves designing, building, and maintaining the infrastructure that people use every day.

"There are as many different types of engineer as you can imagine, but they're all solving people's problems," says Major Rob Ridley, a civil engineer with the British Army's Royal Engineers. "From ensuring the basics, like plenty of clean water in our taps to designing and building the places we live, work and play in, all over the world."[12] With each project, civil engineers work to improve communities and daily life worldwide.

How Do You Become a Civil Engineer?

For Liz Chen, a senior civil engineering student at the University of Notre Dame, building things has been a passion since childhood. "I have always been interested in building structures in which people live and thrive. As a kid, I loved building little houses and buildings with LEGOs," she says. Chen chose to pursue a career that incorporated her love of building and creating structures. "The interest in building structures never faded, and that is why I am majoring in civil engineering now,"[13] she says.

Engineering Education

The first step toward becoming a civil engineer is earning a bachelor's degree in civil engineering or a related field. A bachelor's degree is typically the minimum education required for civil engineers. To earn this degree, students should enroll in a four-year civil engineering program at a college or university.

Many students choose engineering programs accredited by ABET (formerly known as the Accreditation Board for Engineering and Technology). ABET accreditation shows that a school's engineering, technology, or applied science program meets globally recognized high-level quality and rigorous standards. Programs that have earned ABET accreditation have a curriculum that prepares students with the knowledge and skills they will need for their future careers.

Jennifer Stephan is a college admissions counselor based in Massachusetts. She recommends ABET-accredited programs for students interested in civil engineering. Stephan says:

> For some disciplines and careers, think civil engineering and bridge inspection, the ABET-accredited degree is paramount. ABET plays an essential role in licensing professional engineers (PEs), with many states requiring that an engineer have an ABET degree to be licensed. You must be a PE to sign and seal engineering drawings, be in charge of an engineering firm in private practice, or serve as a fully qualified expert witness. Students with these career goals should pursue an ABET-accredited degree.[14]

Common Classes

Students in a civil engineering program typically take a mix of core and elective classes in math, physics, and engineering sciences. Core classes in advanced math, calculus, physics, and engineering mechanics provide a foundational knowledge of engineering. Students also study key areas such as structural engineering, transportation systems, surveying, hydraulics, and construction management. Many students specialize in a specific field of civil engineering, such as structural or geotechnical engineering, and take related elective classes. Civil engineering classes often incorporate hands-on learning through lab work and design projects to apply classroom learning to real-world situations. Students also complete general requirements that vary by school in a broad range of subjects, such as humanities, social sciences, and computer science.

Finn O'Reilly, a 2024 graduate from the University of Notre Dame, majored in civil engineering with a concentration in structural engineering. One of his favorite classes in the civil engineering program was reinforced concrete design. He says:

> I loved the class because we learned how to apply all the fundamental engineering principles that we had been learning since freshman year in order to actually design the structural members that compose buildings and bridges. Moreover, the class involved a semester-long project in which we each individually performed the preliminary, partial design of a 5-story office building. I found this project to be very rewarding because it was a culmination of all the design principles we had learned in class, and it also really sparked my interest in the design element of civil engineering.[15]

Some civil engineers choose to continue their education by earning a master's degree or even a PhD. A graduate degree can open doors to research, teaching, or high-level management positions. However, a bachelor's degree is acceptable for most entry-level jobs.

Internship and Co-Ops

Many civil engineering programs encourage or require students to participate in internships or cooperative education programs, also called co-ops. Working under experienced engineers in an internship or co-op helps students understand how classroom knowledge applies to actual projects.

While both provide real-world experience for students, there are some differences. The primary difference is the length of time. Internships are typically shorter work experiences. They typically occur during several weeks over the summer. Many civil engineering students complete multiple internships while in school to gain experience across various aspects of the field. For example, a

"I loved the class because we learned how to apply all the fundamental engineering principles that we had been learning since freshman year in order to actually design the structural members that compose buildings and bridges."[15]

—Finn O'Reilly, 2024 University of Notre Dame graduate

student may complete one internship in construction engineering and another in geotechnical engineering.

Co-ops also provide valuable work experience but are usually longer than internships. They often extend for a semester or more during the school year. Some schools offer a five-year civil engineering program that includes multiple co-op opportunities.

Audrey Mobed, a student at North Carolina State University, worked a nine-week summer internship at Atlas Engineering, a company that does roof and building restoration and repair work in Raleigh, North Carolina. "I had the opportunity to go on site visits with every engineer before the summer was out. . . . I observed waterproofing work, roof replacements, new K–12 school construction, and residential damage projects,"[16] says Mobed. During the internship, Mobed spent most of her time on a project to restore and repair the Love Building at nearby Davidson–Davie Community College in Thomasville, North Carolina. The building held multiple campus offices, including a library and health services office, but it had experienced severe brick veneer cracking from exposure to sunlight and heat. The project helped Mobed see firsthand real-world applications of several concepts she had learned in school. "At the start of summer, the work Atlas performed on the Love Building had been completely foreign to me. But by the end, I'd

Architecture vs. Civil Engineering

Architecture and civil engineering are closely related fields, but they focus on different aspects of building design and construction. Architects focus on the look, style, and function of a structure. They design buildings to be visually appealing, comfortable, and useful for the people who will use them. Civil engineers, in contrast, focus on making those designs safe and stable. They handle the technical details, such as materials, structure, and foundations, to ensure that the building stands strong and meets safety standards.

Architecture and civil engineering follow different educational paths. Architecture programs focus on creativity, design, and art, as well as technical knowledge. Students study subjects like architectural design, drawing, history, and CAD. Most architecture programs take five years to complete, and students earn a bachelor of architecture degree. In comparison, civil engineering programs focus more on math, physics, and problem-solving and usually lead to a bachelor's degree in four years.

gained a greater understanding of the materials used, construction processes, and the role of the designer on such projects—ideas that are hard to teach in just a classroom setting,"[17] says Mobed.

Licenses and Certifications

More than a century ago, engineers did not need to be licensed. Anyone could work as an engineer without proving their abilities. To protect the public's health and safety, Wyoming passed the first law requiring engineers to be licensed in 1907. Today every state requires engineers who provide services to the public to be licensed. Only licensed engineers are able to sign and seal engineering plans to certify that they directly prepared or supervised the preparation of the plans.

The path to becoming a PE has several steps. After earning a degree from an accredited engineering program, graduates must pass the Fundamentals of Engineering exam, which allows them to become an Engineer in Training. This six-hour exam tests a candidate's knowledge of topics across their discipline, such as civil, mechanical, or electrical engineering. Then they gain about four years of work experience under the supervision of a licensed engineer. Candidates must also pass a second, nine-hour exam, the Professional Engineering exam. Once each step is completed, candidates apply for the Professional Engineer (PE) license with their state licensing board.

Engineers who hold a PE license are known as Professional Engineers (PEs). The PE license shows that an engineer is qualified to take legal responsibility for engineering work and can sign off on projects. There are PEs in most engineering fields, from civil to electrical. A PE can oversee the work of other junior engineers, approve design plans, sign off on projects, and provide services directly to the public. Most states require PEs to take continuing education classes throughout their career to maintain their PE license. While not every civil engineer becomes licensed, having a PE credential can lead to higher salaries, greater responsibility, and more career options.

With assistance from their teacher, two students build a robotic vehicle. Robotics projects encourage teamwork and creativity while giving hands-on experience with engineering concepts.

Many civil engineers choose to earn additional certifications and licenses to demonstrate skills and competencies. Professional organizations issue certifications to engineers who meet specific educational and experience requirements. Some popular certifications include Project Management Professional, Certified Construction Manager, Structural Engineer, and Geographic Information Systems Professional.

First Steps to a Civil Engineering Career

Preparing for a career in civil engineering can start long before college. In high school, students can build the knowledge and skills that engineers use every day. Taking the right classes, joining activities, and exploring the field early can help students get a strong head start in a civil engineering career.

One of the best ways to prepare for a career in civil engineering is to take advanced math and science classes. Civil engineers use math and science to design structures and solve complex problems, so courses like algebra, geometry, trigonometry, physics, and chemistry are beneficial. These subjects teach logical thinking and problem-solving skills that are essential in engineering work. Students who major in civil engineering will be taking many of these classes, so it is a good idea to build a strong foundation in these subjects while in high school.

If a high school offers classes in architecture, drafting, or computer-aided design (CAD), students should take them. These courses teach students how to read and create technical drawings, which are essential in construction and design projects. Learning how to use design software also gives students a helpful skill that engineers often need on the job.

Joining engineering or STEM (science, technology, engineering, and math) clubs and volunteering can also help students learn more about engineering. Competitions such as bridge building or robotics encourage teamwork and creativity while giving hands-on experience with engineering concepts. Helping at a local science museum, tutoring younger students in math, or assisting at a STEM camp can strengthen communication and leadership

Surveying Careers

Some people who earn a degree in civil engineering choose to work as surveyors. A surveyor is a professional who measures and maps the earth's surface to determine property boundaries, land features, and construction layouts. Surveyors use tools like the Global Positioning System, drones, and laser scanners to collect precise data about land elevations, distances, and angles. Their work is essential for planning construction projects, designing roads, and determining legal property lines. Surveyors often work with civil engineers, architects, and city planners to ensure that projects are built in the correct location and meet legal and safety standards.

A career in surveying requires strong math, science, and computer skills. Most surveyors earn a degree in surveying, geomatics, or civil engineering. They must become licensed to work independently. For people who enjoy working both outdoors and with technology, a career in surveying can be rewarding.

"Get curious about the world around you and start to notice how things are designed—from the roads to the buildings to the coastline. Think about the role that infrastructure plays in your life, and how you might want to use your skills to improve the lives of others."[18]

—Alex Bigazzi, civil engineering professor at the University of British Columbia

skills. These experiences also show a commitment to learning and sharing knowledge.

Job shadowing can offer a real-world view of what civil engineers do every day. Spending time with Professional Engineers helps students understand the different areas of civil engineering, from transportation to environmental design. By challenging themselves in school, getting involved in related activities, and exploring real engineering work, high school students can prepare well for a successful career in civil engineering.

For students considering a career in civil engineering, Alex Bigazzi, a civil engineering professor at the University of British Columbia, recommends approaching the built world with curiosity. "Get curious about the world around you and start to notice how things are designed—from the roads to the buildings to the coastline," he says. "Think about the role that infrastructure plays in your life, and how you might want to use your skills to improve the lives of others."[18]

CHAPTER THREE

What Skills and Personal Qualities Matter Most—and Why?

Successful civil engineers use a combination of technical and soft skills to design, build, and maintain infrastructure projects. In addition to having a strong background in math and physics, civil engineers need both strong engineering skills and people skills. Developing and strengthening these abilities can help a person succeed in this career.

Technical Skills

Skills and abilities can typically be divided into two types: technical skills and soft skills. Technical skills, also known as hard skills, are the skills and knowledge a person needs to perform their job. Technical skills are usually learned in school and during hands-on experiences in the field. For a civil engineer, these are the skills needed to take an idea and turn it into a real-life structure.

Civil engineers need a strong foundation in advanced math and science. They need to be able to use physics, algebra, calculus, and geometry to calculate loads and stresses on a building. They use math and science to determine how much weight a bridge can hold or how water flows through a drainage system. Civil engineers use their math and science skills to make project decisions, design projects, and ensure the safety of their projects.

Civil engineers also need an in-depth knowledge of engineering principles. They use their understanding of structures, materials, geotechnics, hydraulics, mechanics, and other engineering concepts to guide decisions and problem-solving during a project.

Modern civil engineers rely on specialized software to do their jobs. AutoCAD software allows engineers to draft detailed designs, while Revit can be used for building information modeling. Civil 3D is another software tool that can be used for designing roads, grading surfaces, and planning utilities. Knowing how to use these software tools can make a civil engineer's work more efficient and accurate.

Civil engineers spend a lot of time in the field at construction sites, and understanding the construction industry is essential. Understanding building materials, safety procedures, building codes, and regulations is part of making sure a project is safe and meets all laws and regulations.

Surveying is another useful technical skill for civil engineers. Surveying is the process of measuring land and understanding its contours before a project's design begins. Strong surveying skills can help an engineer map a potential site and determine where to place a structure.

Project Management and Organizational Skills

While it is essential to have strong technical knowledge in math, science, and engineering principles, success in this field also depends on a variety of soft skills. These are personal qualities that help civil engineers work effectively with others, stay organized, and make good decisions under pressure. Some of the most important soft skills for civil engineers include project management and organizational skills, communication skills, interpersonal skills, problem-solving skills, and decision-making skills.

Civil engineers often oversee multiple projects at the same time, each with its own schedule, budget, and team members. Strong project management and organizational skills help them stay on top of deadlines and ensure that all aspects of a proj-

Computer-Aided Design

Civil engineers use computer-aided design (CAD) software to create detailed drawings and models of structures before construction begins. CAD software allows engineers to design roads, bridges, buildings, and other infrastructure with high accuracy. It helps them visualize how different parts of a project will fit together and identify potential problems early in the design process. With CAD, civil engineers can easily modify designs, test different materials or layouts, and generate 2-D plans or 3-D models. These models can also be shared with architects, contractors, and clients. CAD software has made the design process more efficient by reducing human error, saving time, and ensuring that designs meet safety standards and project requirements.

ect run smoothly. These skills involve planning each stage of the work, setting priorities, and allocating resources such as materials, equipment, and labor.

For example, a civil engineer working on a new highway project must coordinate with surveyors, contractors, and local government officials while making sure construction stays within budget. Using tools like project management software or spreadsheets helps them track progress and make adjustments when needed. Being well organized allows engineers to manage unexpected delays or changes without letting a project fall behind schedule. In addition, having clear systems for documentation and communication ensures that important details are not overlooked.

Civil engineer Janet Benefo has worked in the field for more than twenty years and explains the importance of being organized as a civil engineer:

> When I'm given a brief for a project, the first thing I do, after identifying what is required, is to list the various items/activities and organize those items into a sequence with timeframes to create a program of delivery. This skill helped me to become an effective manager, as I was able to identify the extent of the work on a project and the resources needed. And learning how to carefully monitor the work, I'm able to make sure projects don't overrun or go over budget.[19]

Communication Skills

Communication is another essential soft skill for civil engineers. They must be able to explain complex technical information in ways that different audiences can understand. They need to write clear reports, give presentations, and speak with clients, team members, and stakeholders from the community.

Civil engineers often need to share information with people who may not have an engineering background, such as city officials or the general public. For example, when discussing a new bridge design, an engineer must explain how the project will improve transportation safety without using overly technical terms. Written communication is just as important, and engineers must prepare detailed proposals, environmental impact reports, and

While visiting a job site, civil engineers might need to put their surveying skills to use. Strong surveying skills can help an engineer map a potential site and determine where to place a structure.

construction documents that are accurate and easy to follow. Good communication helps prevent misunderstandings and ensures everyone involved in the project has the information they need.

"Engineers need to be able to communicate effectively. . . . It is important to have knowledge of the project, but it is equally important to have good communication skills."[20]

—Shaukat Syed, senior civil and environmental engineer

Shaukat Syed, a senior civil and environmental engineer for the Georgia Department of Natural Resources, explains that good communication skills are essential for civil engineers:

> Civil engineers do not just sit in cubes all day designing or analyzing mathematical models. They must explain or defend their work. It may be within their own teams, at clients' offices, in public settings, or before regulatory bodies. For the greatest impact in such settings, engineers need to be able to communicate effectively. They need to be able to discuss projects clearly, clarify the reasons for material selection or layout, and explain their methods. They must be able to describe the technical details to reach an agreement with clients or officials. . . . It is important to have knowledge of the project, but it is equally important to have good communication skills.[20]

Interpersonal Skills

Civil engineers rarely work alone. They lead teams of architects, construction managers, technicians, and laborers. Strong interpersonal skills help engineers build positive relationships with these groups and promote teamwork. They are respectful, listen carefully to others, and provide constructive feedback. "A civil engineer isn't just a person who sits behind a [computer] doing calculations or designing all day, but a person who forges relationships, motivates others and talks openly about worldly

"A civil engineer isn't just a person who sits behind a [computer] doing calculations or designing all day, but a person who forges relationships, motivates others and talks openly about worldly issues."[21]

—Ashkan Amiri, civil engineer

issues,"[21] says Ashkan Amiri, a civil engineer with Kier Group, a British construction and civil engineering firm.

Leadership also plays a significant role in interpersonal success. A civil engineer managing a large construction project must motivate team members, resolve conflicts, and maintain a safe, productive work environment. Engineers who are approachable and cooperative help create a stronger, more efficient team. These relationships can also lead to better collaboration across different disciplines, ensuring that projects meet technical, environmental, and community standards. "The varied nature of the profession means that it requires people with different skillsets and experiences to work together,"[22] says Jamie Arnott, a civil engineer at Wallace Stone LLP, a UK engineering consulting firm.

Problem-Solving Skills

Every construction project faces challenges, whether due to design flaws, unexpected weather, or supply-chain issues. Civil engineers must be skilled problem-solvers who can think critically and creatively to find practical solutions when challenges arise. They might analyze data, review plans, and collaborate with experts from other fields to solve problems.

For example, if soil tests reveal that the ground under a planned building is unstable, a civil engineer must quickly develop a new foundation design or find a safer location. Problem-solving also means learning from past projects and applying that knowledge to avoid future issues. Engineers who stay calm under pressure and approach problems logically can keep projects moving forward efficiently.

Decision-Making Skills

Civil engineers constantly make decisions that affect public safety, project costs, and environmental impact. They must balance many factors simultaneously, including technical feasibility, budget constraints, deadlines, and local regulations. Good decision-making requires gathering accurate information, weighing options carefully, and considering both short- and long-term consequences.

For example, choosing between different construction materials involves evaluating the strength, cost, availability, and sustainability of each option. Civil engineers must be confident in their choices and justify them to clients or regulators. Sound decision-making helps ensure that projects are successful, safe, and cost-effective.

To succeed, civil engineers need to combine solid technical skills with strong soft skills. The ability to organize complex projects, communicate clearly, collaborate with others, solve problems, and make informed decisions allows civil engineers to bring their designs to life efficiently and responsibly. Developing these skills helps engineers adapt to challenges and lead projects that improve the world.

What Is It like to Work as a Civil Engineer?

A typical day for a civil engineer can vary widely depending on the person's specialty, employer, and current project. One day, a civil engineer might be in an office, reviewing computer-aided designs. The next day, they might travel to a construction site to evaluate how a project is progressing. Some projects require civil engineers to be on-site, so they might need to set up a temporary office, often in a trailer, at the construction site.

For many civil engineers, the variety in day-to-day work is one of the reasons they love their career. Jamie Smith is a civil engineer with Ryan Companies, a commercial real estate services firm. She agrees that every day brings something new in her job:

> The best part about my job is probably the versatility of the work. One thing I was afraid of when pursuing engineering was that I would end up sitting behind a desk and doing the same thing every single day. At Ryan, we are looking at so many different sites for so many different types of buildings; every day is different. One day I could be working on a pursuit for a medical office building in Minnesota, and the next day working on a senior living facility in Florida. Each site also comes with its own set of challenges, such as wetlands, floodplains, soil conditions, topography, and local code requirements. Then you are trying to balance the site conditions with the size of the building and other client needs. It is like one giant puzzle, and I love puzzles.[23]

A Day in the Life of a Structural Engineer

Joan Lim is a civil engineer who specializes in structures, also called a structural engineer. She works for Mott MacDonald, a global engineering consulting firm, in Singapore. As a child, Lim was fascinated with puzzles and building. As she grew older, these interests turned into an appreciation for structures that improve daily life. In school, Lim chose to pursue civil engineering so she could build a career designing structures that would improve communities. "I realized that structural engineering combines creativity, problem-solving, and the opportunity to create something tangible that impacts people's lives,"[24] she says.

Lim starts out by reviewing tasks and projects to determine her priorities for the day, making sure any urgent requests from clients, contractors, or her team are handled. She spends part of the day working with engineering software to create the framework of a building design. She carefully calculates each beam, column, and foundation in the design. While working, Lim meets with the project's architects to make sure her design incorporates the aesthetics and functionality they envisioned. She also meets with contractors to ensure that their construction methods match the plans and that the project meets expectations and safety standards.

Sometimes, Lim leaves the office and travels to project sites in the field. During these site visits, she inspects the project's progress and addresses any problems or challenges that arise. Sometimes, unexpected site conditions can create challenges. For example, on one project, Lim and her team had to get creative with a tight timeline when traditional construction methods could not meet the client's deadlines. Lim's team adjusted their design to use prefabricated and precast elements, which significantly reduced the time needed on-site for construction. She worked closely with the project's contractors and local regulators to make sure the project stayed on schedule. "It was a true test of teamwork, creativity, and resilience," she says. "It showed how embracing innovation and working collaboratively can overcome even the toughest challenges."[25]

Working in Water Resources

Riley Mondloch is a civil engineer who specializes in water resources. In his job, Mondloch frequently works on flood management and stormwater systems. Most of the time, he works on stormwater management projects in Minnesota and Wisconsin, but occasionally he travels to Colorado or Texas for projects. In a typical day, Mondloch spends most of his time working at a computer, preparing flood models, and writing reports. "While I do venture into the field occasionally for data collection or troubleshooting, much of my work is conducted remotely, using technology to streamline processes and maximize efficiency," he says.

Mondloch enjoys working in water resources because it allows him to use his creativity and problem-solving skills. "There are many flooding issues to solve, and since I really enjoy problem-solving and figuring things out, it's a great fit," he says. Sometimes finding a solution to a flooding problem can be complicated, especially in well-developed areas, where engineers must work around existing infrastructure. However, Mondloch enjoys the challenge. "It allows me to think outside the box, challenge myself, and get more creative," he says.

Quoted in Short Elliott Hendrickson, "Day in the Life of a Water Resources Engineer," May 13, 2024. www.sehinc.com.

Lim says the best part of her day is seeing one of her designs become a real-world structure. "The most rewarding part is watching something I've worked on transform from an idea into a physical space that people will use and experience," she says. "There is nothing quite like standing inside a building and knowing you had a hand in making it possible."[26]

Who Hires Civil Engineers?

Civil engineers work for a variety of employers. More than half (52 percent) of all civil engineers work for engineering services firms, according to the Bureau of Labor Statistics' *Occupational Outlook Handbook*. Civil engineering services firms are companies that provide the planning, design, and management of construction projects for infrastructure and buildings. They work on public and private projects, including roads, bridges, and water systems, and offer a range of services like site development, environmental assessments, and structural analysis. These firms may specialize in certain areas, such as transportation or water management, or offer a wide range of services

Some civil engineers work for state (11 percent), local (10 percent), or federal (3 percent) governments, according to the *Occupational Outlook Handbook*. In these roles, they plan, design, and supervise public infrastructure projects such as roads, bridges, and water systems, while ensuring compliance with regulations. Their work is essential for community development, with many jobs involving both office-based design and site-based oversight.

Civil engineers working for nonresidential building construction companies (7 percent) are involved in the planning, design, and oversight of large-scale projects like offices, factories, and hospitals. Their responsibilities include analyzing site data, using design software for structural and systems planning, and ensuring compliance with regulations. They also monitor construction progress on-site to ensure safety and quality, checking details such as the strength of foundations and structural frameworks before concrete is poured.

> **"The most rewarding part is watching something I've worked on transform from an idea into a physical space that people will use and experience. There is nothing quite like standing inside a building and knowing you had a hand in making it possible."[26]**
>
> —Joan Lim, civil engineer

Where Do Civil Engineers Work?

Civil engineers divide their time between office work and site work. In the office, they design structures, create plans using CAD software, analyze data, prepare reports, and communicate with clients or government officials. In-office work typically focuses on planning, design, and coordination.

At construction sites, civil engineers oversee projects, inspect materials, ensure that safety standards are met, and address real-time problems that arise during construction. Site work is more hands-on and requires teamwork with contractors and laborers. Both settings are essential in a civil engineering job. Office work ensures accuracy and compliance, while site work ensures that the project is built correctly and safely.

Anna Prescott is a geotechnical engineer with CGL, a geotechnical consulting firm in the United Kingdom. She spends about half of her time in the office and the other half in the field. Prescott writes proposals for new projects, talks to subcontractors and clients, and writes reports. Reports range from simple desk studies, which might take a day to write, to complex interpretative reports, including in-depth analysis of the ground encountered. "I have also been involved in modeling projects, where we analyze the potential movements of the ground,"[27] she says.

> **"Site works are always well-planned, but nobody can foresee the unpredictable nature of the ground beneath our feet! As a result, I have to be ready to solve any problem which arises, such as finding concrete obstructions or navigating an unexpected steep slope."[28]**
>
> **—Anna Prescott, geotechnical engineer**

In the field, Prescott visits clients' sites to supervise the drilling of boreholes, the excavation of trial pits, or the collection of soil and rock samples for testing. "Site works are always well-planned, but nobody can foresee the unpredictable nature of the ground beneath our feet!" she says. "As a result, I have to be ready to solve any problem which arises, such as finding concrete obstructions or navigating an unexpected steep slope."[28] Prescott recounts one project that involved weeks of site work in very remote areas, only accessible using 4x4 vehicles and long walks.

Work Schedules and Earnings

A civil engineer's work schedule can vary depending on their role and the stage of a project. Most civil engineers work full time and typically put in about forty to fifty hours per week. In the office, engineers usually work regular business hours, such as Monday through Friday, 9:00 a.m. to 5:00 p.m. In the office, they spend their time designing projects, creating plans, and meeting with clients or government officials.

In the field, civil engineers often have more varied schedules. They may start earlier in the day and work longer hours, especially when overseeing construction, inspecting sites, or troubleshooting problems. During major construction phases or project deadlines, civil engineers may work evenings or weekends to keep projects on schedule.

Several factors can cause civil engineers to work additional hours. Project deadlines, such as when a bridge design or a road survey must be completed by a specific date, often require civil engineers to work overtime. Unexpected issues on-site, such as challenging ground conditions or material delays, can cause additional work for civil engineers since they must adjust schedules to get a project back on track. Sometimes, coordinating with

Civil engineers who work for government agencies plan, design, and supervise public infrastructure projects. Roads, bridges, and water systems are among the projects these engineers might work on.

project teams in different time zones or handling client requests at the end of the day causes engineers to work longer to ensure that everything gets done.

According to the Bureau of Labor Statistics' *Occupational Outlook Handbook*, the median annual wage for civil engineers was $99,590 as of May 2024. The top-paying employer was the federal government, with a median annual wage of $114,210.

CHAPTER FIVE

Advancement and Other Job Opportunities

The first job for most civil engineers is an entry-level or junior engineer position. However, with experience and hard work, civil engineers can advance into various specialties and leadership roles within different organizations.

Early Career Opportunities

After college, new civil engineers usually begin as junior or entry-level engineers. In this stage, they work under senior engineers and learn the practical aspects of the job. They may help prepare design drawings, conduct surveys, analyze data, or assist with project reports. Entry-level engineers often visit construction sites to see how projects come together in real life.

As they gain experience, civil engineers can move into more advanced roles. They might become project engineers, responsible for managing parts of a project and ensuring that everything meets technical and safety standards. Others may specialize in specific areas, such as structural, geotechnical, or transportation engineering.

Civil engineers often work for engineering consulting firms, construction companies, or government agencies. Each type of employer offers different experiences, responsibilities, career paths, and opportunities.

Mid-Career Growth and Leadership Roles

With experience, civil engineers often take on leadership and management positions. Most people who advance into management or leadership roles have a graduate degree, such as a master's degree or PhD, and have earned their Professional Engineer (PE) license. Project managers oversee entire construction or design projects, coordinating teams of engineers, architects, and contractors. They manage budgets, timelines, and communication between all parties involved. Other engineers move into supervisory roles, where they guide younger engineers and make key technical decisions. Some become consultants, offering expert advice on specialized topics such as seismic safety or sustainable materials.

Civil engineers with strong communication and organizational skills can also advance into executive or administrative roles, such as city engineer, department head, or company director. In these positions, they are involved in large-scale decisions that shape communities and policies.

Urban Planning

Some civil engineers pursue careers in urban planning. An urban planner is a professional who designs and organizes how cities and communities grow and develop. They create plans for land use, transportation systems, housing, parks, and public spaces to ensure that areas are safe, efficient, and enjoyable for the people who live there. Urban planners study population trends, environmental issues, and economic factors to decide the best ways to use land and resources. They often work with government officials, engineers, and architects to design neighborhoods, road systems, and city centers.

A civil engineer can become an urban planner by building on their engineering background and gaining knowledge in city planning and development. Since civil engineers already understand infrastructure, transportation, and construction, they have a strong foundation for planning how cities grow. They can earn a master's degree or certificate in urban or regional planning, which covers topics such as land use, zoning laws, sustainability, and community design. Civil engineers can also gain experience in urban planning by working on projects related to transportation planning, public works, or environmental design. Joining professional organizations such as the American Planning Association can help them network and learn more about opportunities.

Mid-career engineers might also switch paths to work in research, education, or government policy. Teaching at universities or working on infrastructure planning for public agencies allows engineers to share knowledge and shape the future of the profession. Hans Tritico is a professor of civil engineering at the University of Mount Union in Ohio. Tritico holds a dual PhD in civil engineering and aquatic ecology from the University of Michigan and has spent more than a decade specializing in water resources. Tritico uses project-based learning in his engineering classrooms at Mount Union. In their projects, students gain hands-on experience helping real clients in the community. "Pretty early on in my teaching, I realized that it wasn't enough to simply teach a subject. Rather, my role is to 'grow people.' Building successful, young professionals means caring about—and pushing—each one of my students,"[29] he says.

Skills for Successful Leaders

Successful managers and leaders have a few things in common, according to Shelley Row, a licensed transportation engineer. Row previously held leadership roles as director of the Intelligent Transportation Systems Joint Program Office at the US Department of Transportation and as director of the Office of Transportation Operations at the Federal Highway Administration. She is currently the founder and chief executive officer of Blue Fjord Leaders, a company that provides management and leadership training for engineers and other technical professionals.

Row says that civil engineers need to expand their skill sets to succeed as they move into management and leadership roles. "We engineers tend to [be] detail-oriented, logical, meticulous, rational, data-driven, precise, methodical, and linear," says Row. "We are problem solvers. That's how our brains are wired. And those are strong, desirable traits for managers and leaders, but they aren't the only traits needed to be successful managers."[30]

Row explains that civil engineers who successfully transition to management and leadership roles display three common soft skills: interpersonal communication, the ability to develop relationships, and self-awareness. She says:

> If you think about it, all three skills speak to how a manager does their job, not what they do. For example, no one I interviewed said their favorite manager was a great delegator or a great decision-maker or good at performance reviews. They focused on how they approach their management role. Once you master those three foundational "how" skills, you can use them to be a better delegator or a better mentor or to handle a difficult conversation or other management functions.[31]

The Importance of Networking

In civil engineering, like many fields, building strong professional connections is essential for success. Networking opens doors to new opportunities, allows engineers to share knowledge, and connects civil engineers with mentors who can provide valuable guidance. Networking can provide a person access to advice, opportunities, and resources they might not otherwise have. Whether an engineer is in their first job or is well established in the field, effective networking can be a powerful tool throughout one's career.

Building a professional network in civil engineering can seem challenging at first, but it usually begins with small, simple steps. A strong network often starts with familiar connections such as classmates, coworkers, and acquaintances in the industry. These early relationships form the base of a professional network. "Networking within your organization is a game-changer for career advancement," says Shahab Naveed Rashid, a licensed civil engineer. "Building strong relationships with colleagues across departments not only opens up new opportunities but also fosters collaboration and knowledge sharing. It's all about leveraging internal connections to grow professionally and contribute more effectively to your projects."[32]

"Networking within your organization is a game-changer for career advancement. Building strong relationships with colleagues across departments not only opens up new opportunities but also fosters collaboration and knowledge sharing."[32]

—Shahab Naveed Rashid, licensed civil engineer

Civil engineers can expand their networks by attending industry events such as conferences, seminars, and workshops. These events are an effective way to meet other engineering professionals, exchange ideas, and learn about new developments in civil engineering. Joining professional organizations, such as the American Society of Civil Engineers (ASCE), also provides valuable networking opportunities through events, publications, and online communities.

In today's digital world, online networking tools are increasingly important for career development. LinkedIn is one of the most effective platforms, allowing professionals to create detailed profiles, join civil engineering groups, and participate in discussions with others within the field. Social media platforms like X (formerly Twitter) also provide spaces for engineers to follow industry leaders, share insights, and take part in conversations about the latest innovations and trends. These online tools make it easier to connect with professionals across different regions and specializations.

For James Hall, the mentors he found through networking have had a positive impact on his career. Hall is a civil engineer and began his career as an assistant project engineer for Network Rail in the United Kingdom. Today he is the associate director, engineering manager for WSP. He says:

> Without doubt, the mentorship I've had at various points has helped me to successfully progress throughout my career. When I started out as a placement student with

Network Rail, the senior engineer took me under their wing. They introduced me to a library of standards and processes, and taught me about different railway sub-systems, their interactions and interfaces. As a mentor, they have always offered advice, whether it's a technical problem or career progression. They're the reason I've been able to develop as a contractor's engineering manager and leader.[33]

> **"Without doubt, the mentorship I've had at various points has helped me to successfully progress throughout my career."[33]**
>
> **—James Hall, civil engineer**

A career in civil engineering offers many opportunities to grow and advance. Civil engineers can focus on a specific field, take on leadership roles or management positions, or even work in different industries. With ongoing learning and strong professional connections, civil engineers can discover new opportunities and reach their full career potential.

What Does the Future Hold for Civil Engineers?

The demand for civil engineers will remain strong in the future because society will always need infrastructure. Roads, bridges, and water systems wear out over time and must be replaced or repaired. In addition, population growth and climate change are creating new challenges that require innovative engineering solutions. "One of the great things about being a civil engineer is that it is quite resilient to economic cycles," says Alex Bigazzi, a civil engineering professor at the University of British Columbia. "When the economy is doing well, there are lots of large projects underway. When the economy is struggling, governments will often invest in infrastructure projects to provide economic stimulus. That means there is a steady demand for civil engineers, rather than the spikes and drops that we might see in other areas."[34]

> "One of the great things about being a civil engineer is that it is quite resilient to economic cycles."[34]
>
> —Alex Bigazzi, civil engineering professor at the University of British Columbia

Job Growth

According to the Bureau of Labor Statistics, employment of civil engineers is expected to grow steadily over the next decade, at a rate of about 5 percent through 2034. This growth is faster than the average rate for all occupations. An estimated 23,600 openings are projected each year over the decade, driven by both new job creation and the need to replace workers.

The projected future job growth for civil engineers is driven by the need to maintain aging infrastructure and build new, sustainable, and resilient cities. A significant need exists for civil engineers to manage projects to rebuild and reinforce aging roads, bridges, dams, and other critical infrastructure that have passed their original lifespans.

In March 2025 the ASCE issued its *2025 Report Card for America's Infrastructure*. Issued every four years, the report card examined infrastructure nationwide, including roads, dams, drinking water, and railroads. The ASCE gave the country's infrastructure a C grade, up slightly from previous reports, but still concerning. ASCE engineers warned that ignoring aging infrastructure would lead to bigger problems in the future. "As the Report Card makes clear, there is still a great need for sustained investments," says Feniosky Peña-Mora, 2025 ASCE president. "Delaying upgrades to our nation's roads, bridges, transit, and utilities will cost families and businesses time and money, in addition to creating unsafe yet often avoidable situations."[35]

Beyond just maintenance and repair of aging infrastructure, civil engineers will also be needed to design and build new public works that are crucial for modern society. Civil engineers will be needed for projects to improve transportation networks, such as new roads and tunnels, and structures designed for public safety and resilience, such as flood barriers and earthquake-proof buildings. As air travel expands, there is a significant need to design and renovate airports to handle more passengers and cargo, and civil engineers will be needed to manage these projects.

Emerging Trends: Climate-Resistant and Sustainable Designs

The future of civil engineering includes the creation of climate-resistant and sustainable designs. As the effects of climate change become more severe, engineers are rethinking how

Civil engineers use drones to survey large areas quickly and safely, capturing detailed images and measurements. Drones help engineers track progress and identify issues on construction sites in real time.

to build infrastructure that can handle extreme weather, rising sea levels, and changing environmental conditions. Climate-resistant designs include stronger bridges, flood barriers, and buildings that can withstand hurricanes, floods, and heat waves. Civil engineers are also working to incorporate materials in their projects that last longer and reduce carbon emissions, such as recycled concrete and low-carbon steel.

Sustainability is another key trend for the future of civil engineering. Civil engineers are designing green buildings that use renewable energy sources such as solar and wind power. They are designing systems to reuse water, reduce waste, and protect natural resources. Smart cities, which use technology to manage energy and transportation efficiently, are another part of this sustainable future.

New Technologies in Civil Engineering

Advancements in technology have transformed the field of civil engineering, making it more precise, efficient, and innovative than ever before. Modern engineers rely on digital tools and smart systems to design, plan, and maintain infrastructure projects. Technologies such as Building Information Modeling (BIM), geographic information systems, drones, and artificial intelligence (AI) are changing the way engineers approach construction and problem-solving.

BIM is one of the most important tools in use today for construction and infrastructure projects. BIM allows engineers to create detailed 3-D digital models of buildings, bridges, and other structures. Unlike traditional blueprints, BIM models include the structure's shape and size, along with important information about materials, costs, schedules, and how different systems, such as plumbing or electricity, work together. With BIM, civil engineers can visualize projects before construction begins, detect design problems early, and improve communication among architects, engineers, and contractors. BIM helps save time, reduce errors, and lower costs. BIM also supports sustainability by allowing engineers to analyze energy use and environmental impact. As

Emerging Roles for Civil Engineers

New roles for civil engineers are emerging as technology and sustainability solutions become more important. These roles are changing how engineers approach challenges and plan for the future. For example, one growing role is a Building Information Modeling (BIM) specialist. A BIM model is a digital 3D representation of a building or infrastructure project that includes data about every part of the structure. BIM specialists create and manage these models and collaborate with architects, other engineers, and contractors throughout a project. Another role is a sustainability consultant, who helps construction projects reduce their impact on the environment. These consultants guide companies in using renewable materials, lowering energy consumption, and designing eco-friendly buildings. Their goal is to create structures that are durable while protecting the planet. These emerging roles in civil engineering demonstrate the demand for people who can use technology to create smart, sustainable solutions that improve everyday life in communities worldwide.

construction projects become more complex, understanding BIM gives civil engineers the skills needed to plan, design, and manage projects efficiently and effectively, making it an important part of their professional skill set.

Alex is a civil engineer who relies on BIM software for his projects in the construction industry. In a blog post for Modelo.io, an online 3D visualization and collaboration platform for architects, engineers, and construction teams, Alex writes:

> Using BIM software, I can visualize every detail of a construction project, from the placement of electrical systems to the layout of interior spaces. This not only helps our clients better understand the project but also allows us to make informed decisions about design and construction. As a BIM engineer, I also play a key role in ensuring that construction projects are efficient and sustainable. By analyzing data and simulating different scenarios, we can optimize energy usage and reduce environmental impact. This is a significant aspect of our work, as we strive to create buildings that are not only structurally sound but also environmentally friendly.[36]

Geographic information systems help civil engineers analyze spatial and environmental data. This technology is especially valuable when planning transportation systems, water management projects, or city layouts. By using a geographic information system, engineers can study land features, population density, and environmental factors to choose the most suitable and sustainable locations for new infrastructure.

Drones are also being used in civil engineering. Drones are used to survey large areas quickly and safely, capturing detailed images and measurements that were once time-consuming to gather. They help engineers track progress and identify issues on construction sites in real time. Riley Mondloch, a civil engineer who specializes in water resources, uses drones in his work. "Drones have revolutionized surveys and data collection, replacing traditional methods with more efficient alternatives," he says.[37]

Artificial Intelligence

Artificial intelligence is another technology that is revolutionizing civil engineering. AI software can create design ideas, generate project management plans and budgets, and even suggest ways to make structures more sustainable. By using AI tools, civil engineers can improve the accuracy, efficiency, and sustainability of their work.

Civil engineers are incorporating AI into their work to make structures safer. AI-powered sensors and systems constantly monitor the condition of structures such as bridges, tunnels, and buildings. They detect signs of stress, cracks, or damage early, allowing engineers to fix problems before they become serious. On the ground, AI can analyze large amounts of data to predict soil behavior and identify risks such as landslides or unstable ground. These tools help geotechnical engineers choose safe construction sites and plan projects more effectively, especially in difficult environments.

Some civil engineers use generative AI tools to create smarter, more efficient building designs. Engineers can enter project details and let the AI tool generate several design options that meet cost, energy use, and performance requirements. AI tools support sustainable building by suggesting eco-friendly materials and energy-efficient designs. Some AI tools can analyze a building's entire life cycle to minimize waste and energy use.

Generative AI can also be used with what are known as digital twins, which are virtual replicas of real-world structures. Engineers use AI and digital twins to simulate real-world conditions and test how a structure or system will respond to loads, stress, weather, or earthquakes before the new structure is built. With digital twin models, engineers can see how a bridge will handle loads ten years in the future, how a drainage system will respond to extreme

> **"With the help of AI, civil engineers can already design and construct smarter buildings, bridges, and infrastructure."[38]**
>
> —Rizwan Siddiqi, PE, president and chief executive officer of EBA Engineering

weather, or how a building will handle hurricane-force winds. With this information, civil engineers can adapt their designs before construction begins. "With the help of AI, civil engineers can already design and construct smarter buildings, bridges, and infrastructure," says Rizwan Siddiqi, PE, president and chief executive officer of EBA Engineering. "Civil engineers can also optimize their decision-making processes by leveraging AI technologies, improving project efficiency, automating daily processes and reporting requirements, enhancing site safety, and more."[38]

AI and Smart Cities

AI can also help civil engineers design smart cities. These are urban areas that use technology and data to improve the quality of life for residents. They use sensors, cameras, and digital systems to manage services like traffic, energy, waste, and public safety more efficiently. With AI, civil engineers can design infrastructure that operates more efficiently and sustainably, integrating data from traffic systems, energy grids, environmental sensors, and more.

For example, civil engineers can use AI tools to analyze traffic flow and congestion patterns in cities. With this data, they can optimize traffic light timing, improve road layouts, and predict traffic problems before they happen. These improvements help reduce delays, save fuel, and create safer, more efficient transportation systems. "Technology is the connective tissue between every project in a smart city, and it requires a lot of small projects connected together. They become the digital infrastructure. You can't necessarily see them, but they make the quality of daily life better," says Shaurya Agarwal, an associate professor in civil engineering at the University of Central Florida.[39]

Overall, these advanced technologies are creating a smarter, safer, and more efficient future for civil engineering. They enable engineers to make better decisions, minimize waste, and design infrastructure that meets the growing demands of society. As technology continues to evolve, civil engineers who understand and use technology will be in demand.

The Future of Civil Engineering

The future of civil engineering looks bright and full of possibilities. As technology advances, engineers are using artificial intelligence, drones, and 3-D modeling to design smarter and safer infrastructure. Civil engineers are also working on sustainable designs and structures that can reduce carbon emissions and withstand extreme weather events. Civil engineers will continue to have an essential role in creating the communities where people can live and thrive in a changing world.

SOURCE NOTES

Introduction: Rebuilding the Fern Hollow Bridge

1. Quoted in Kris B. Mamula et al., "Witnesses Describe What They Heard and Saw When Pittsburgh Bridge Collapsed," *Pittsburgh Post-Gazette*, January 28, 2022. www.post-gazette.com.
2. Jason Fuller, "From the Field: How Pittsburgh's Fern Hollow Bridge Was Replaced in Less than a Year," American Society of Civil Engineers, 2024. www.asce.org.
3. Fuller, "From the Field."
4. Fuller, "From the Field."
5. Quoted in Christopher DeRose, "Fern Hollow Bridge Reopens After Final Touches," CBS News, July 8, 2023. www.cbsnews .com.

Chapter One: What Does a Civil Engineer Do?

6. Quoted in Ross Group, "A Day in the Life of a Structural Engineer," January 26, 2021. www.withrossgroup.com.
7. Quoted in Ross Group, "A Day in the Life of a Structural Engineer."
8. Quoted in Ross Group, "A Day in the Life of a Structural Engineer."
9. Quoted in Rebecca McElhatten, "Kleinfelder Plays Key Role in Opening of San Diego International Airport's New Terminal 1," Kleinfelder, October 13, 2025. www.kleinfelder.com.
10. Melinda Eason, "10 Reasons to Pursue a Career in Geotechnical Engineering," LinkedIn, April 7, 2025. www.linkedin.com.
11. Kelley Kircher, "Reflecting on a Career as a Transportation Engineer," LinkedIn, November 17, 2022. www.linkedin.com.
12. Quoted in Anh Nguyen, "7 Top Reasons Why Civil Engineering Is the Best Career," Institution of Civil Engineers, 2023. www .ice.org.uk.

Chapter Two: How Do You Become a Civil Engineer?

13. Quoted in Shannon Rooney, "Considering Civil Engineering? Three Students on Why They Chose the Major and What They Love About It," Undergraduate Admissions, University of Notre Dame, August 23, 2023. https://admissions.nd.edu.

14. Jennifer Stephan, "For Aspiring Engineers: Understanding the Value of ABET Accreditation," Lantern College Counseling, November 2, 2022. www.lanterncollegecounseling.com.
15. Quoted in Rooney, "Considering Civil Engineering?"
16. Ashley Mobed, "My Eventful, Insightful Summer at a North Carolina Engineering Firm," American Society of Civil Engineers, 2025. www.asce.org.
17. Mobed, "My Eventful, Insightful Summer at a North Carolina Engineering Firm."
18. Quoted in UBC Engineering, "Why Study Civil Engineering?," 2024. https://engineering.ubc.ca.

Chapter Three: What Skills and Personal Qualities Matter Most—and Why?

19. Janet Benefo, "5 Transferable Skills That Are Important to Civil Engineering," Institution of Civil Engineers, 2021. www.ice.org.uk.
20. Shaukat Syed, "Public Speaking Is a Critical Skill Civil Engineers Need to Learn," American Society of Civil Engineers, 2025. www.asce.org.
21. Quoted in Nguyen, "7 Top Reasons Why Civil Engineering Is the Best Career."
22. Quoted in Nguyen, "7 Top Reasons Why Civil Engineering Is the Best Career."

Chapter Four: What Is It like to Work as a Civil Engineer?

23. Quoted in Ryan Companies, "International Women in Engineering Day: What It's like Being a Female in Civil Engineering," June 23, 2023. www.ryancompanies.com.
24. Quoted in Building and Construction Authority, "A Day in the Life of a Structural Engineer," *Build SG Magazine*, August 26, 2025. www1.bca.gov.sg.
25. Quoted in Building and Construction Authority, "A Day in the Life of a Structural Engineer."
26. Quoted in Building and Construction Authority, "A Day in the Life of a Structural Engineer."
27. Quoted in Erin Martin-Jones, "What Does It Mean to Be a Geotechnical Engineer? Anna Prescott Explains," Earth Sciences, University of Cambridge, January 30, 2023. https://blog.esc.cam.ac.uk.
28. Quoted in Martin-Jones, "What Does It Mean to Be a Geotechnical Engineer?"

Chapter Five: Advancement and Other Job Opportunities

29. Quoted in University of Mount Union, "Dr. Hans Tritico," 2024. www.mountunion.edu.
30. Quoted in American Society of Civil Engineers, "Top Tips for Transitioning from an Engineer to a Manager," September 1, 2021. www.asce.org.
31. Quoted in American Society of Civil Engineers, "Top Tips for Transitioning from an Engineer to a Manager."
32. Quoted in LinkedIn, "Here's How You Can Advance Your Career by Networking Within Your Own Organization as a Civil Engineer," 2024. www.linkedin.com.
33. James Hall, "Why You Shouldn't Underestimate the Power of Mentoring," Institution of Civil Engineers, June 4, 2025. www.ice.org.uk.

Chapter Six: What Does the Future Hold for Civil Engineers?

34. Quoted in UBC Engineering, "Why Study Civil Engineering?"
35. Quoted in American Society of Civil Engineers, "ASCE Report Card Gives U.S. Infrastructure Highest-Ever C Grade," March 25, 2025. www.asce.org.
36. Quoted in Modelo, "Life as a BIM Engineer: A Day in the Construction Industry," June 29, 2024. www.modelo.io.
37. Riley Mondloch, "Day in the Life of a Water Resources Engineer," Short Elliott Hendrickson, May 13, 2024. www.sehinc.com.
38. Rizwan Siddiqi, "How Generative AI Will Transform Civil Engineering," EBA Engineering, June 11, 2024. https://ebaengineering.com.
39. Quoted in Robert Stephens, "What Exactly Is a Smart City—and Why Should All of Us Care," University of Central Florida, September 23, 2024. www.ucf.edu.

INTERVIEW WITH A CIVIL ENGINEER

Jim Venture, PE, is the chief executive officer and managing principal of PVEDI Engineering, a national architecture and engineering firm. He has more than thirty years of experience as a licensed civil engineer. In addition to his leadership role, Venture is still an active member of the company's Civil and Land Development Team, working on commercial, industrial, and residential land development projects and more. He answered questions about his career by email.

Q: How long have you worked as a civil engineer?
A: I graduated from Penn State University in 1993 and have worked as a civil engineer for over 30 years. During that time, I've been fortunate to lead and contribute to a wide range of projects, from residential developments to large-scale commercial and infrastructure work.

Q: Why did you become a civil engineer?
A: From a young age, I was fascinated by how things are built and enjoyed being outdoors. Civil engineering offered the perfect combination of creativity, problem-solving, and hands-on results. I was drawn to the idea of designing something on paper and then seeing it come to life in the real world. There's something incredibly rewarding about knowing that your work can transform land into something useful and lasting.

Q: What does your typical workday look like?
A: As the [chief executive officer], my current role involves overseeing a variety of engineering and design disciplines, including land development, environmental, structural, surveying, and architecture. I spend much of my day coordinating with department directors, reviewing project progress,

and ensuring that our designs meet both client goals and regulatory standards. I also collaborate with our accounting and finance teams on budgeting and operations. Although my responsibilities are broader now, I still stay closely involved with some select projects that require specialized input or creative solutions.

Q: What do you like most about your job?
A: What I enjoy most is seeing ideas take physical form. Watching a site evolve from an open piece of land into a completed project. I also appreciate the diversity of the work as civil engineering offers constant variety; every project brings new challenges, sites, and solutions, which keeps the job engaging and dynamic.

Q: What do you like least about your job?
A: Timesheets. In our profession, we must carefully track how every hour is spent so projects can be billed accurately and managed effectively. It's a necessary part of maintaining professionalism and accountability, even if it's not the most exciting aspect of the job.

Q: What personal qualities do you find most valuable for this type of engineering work?
A: Beyond strong technical and analytical skills, successful civil engineers need excellent communication and interpersonal abilities. Much of our work involves collaboration with clients, government agencies, contractors, and the public, so being able to convey technical information clearly and respectfully is critical. Equally important are problem-solving skills, attention to detail, and the ability to adapt to changing circumstances and challenges.

Q: What is the best way to prepare for this type of engineering job?
A: The best preparation is to focus on math and science courses in high school, as they form the foundation for all engineering disciplines. In college, seek out a program with a strong reputation

for engineering excellence and an active alumni network, as those connections often open doors later in your career. Also, try to gain practical experience through internships or summer jobs related to construction or design to better understand how engineering principles are applied in the real world.

Q: What advice do you have for students who might be interested in a career as a civil engineer?
A: Choose a career that excites you. Civil engineering is ideal for those who enjoy problem solving, teamwork, and seeing tangible results from their efforts. If you're someone who likes to build, design, and improve the spaces where people live and work, this field offers endless opportunities to make a lasting impact on your community. It's a profession that combines creativity with purpose, and that's what makes it so rewarding.

American Society for Engineering Education (ASEE)
www.asee.org
The ASEE is a nonprofit professional organization that works to advance engineering and engineering technology education. Its website features publications, news, articles, and information about webinars and other learning opportunities.

American Society of Civil Engineers (ASCE)
www.asce.org
The ASCE is the oldest national engineering society in the United States. Its website has information for civil engineers, including current publications, news, education, events, and articles about the latest topics impacting the industry.

Engineers for a Sustainable World (ESW)
www.eswglobal.org
ESW is a global nonprofit network of students and professionals, including civil engineers, that share a passion for technical sustainability. Its website has information about local chapters, upcoming events and programs, and recent news articles.

National Society of Professional Engineers (NSPE)
www.nspe.org
The NSPE is a professional association for licensed Professional Engineers, including civil engineers. Its website has information about continuing education, job postings, news, publications, and other resources for engineers.

Society of Women Engineers (SWE)
https://swe.org
The SWE is an international not-for-profit educational and service organization that advocates for women in engineering and technology, including civil engineering. Its website has information about scholarships, upcoming programs and events, and other resources.

INDEX

Note: Boldface page numbers indicate illustrations.

ABET accreditation, 17–18
Agarwal, Shaurya, 51
American Planning Association, 40
American Society of Civil Engineers (ASCE), 43, 46
Amiri, Ashkan, 29–30
architects/architecture, 20
Arnott, Jamie, 30
artificial intelligence (AI), 50–52
AutoCAD software, 26

Bakth, Melissa, 5
base isolation, 14
Benefo, Janet, 27
Bigazzi, Alex, 24, 45
bridges, designing, 6
building information modeling (BIM), 48–49
buildings, 11
Bureau of Labor Statistics
  earnings, 38
  employers, 34, 35
  job outlook, 45–46
Burj Khalifa (Dubai, United Arab Emirates), 11

career ladder, 39–41
certification, 4, 21–22
Chen, Liz, 17
Civil 3D, 26
civil engineers
  basic facts about, 7–10
  process followed by, 10–11
  specialty areas of, 12–15, **13**
communication skills, 28–29, 40
computer-aided design (CAD) software
  capabilities of, 10, 27
  classes in, 23
  types of, 26
computers
  AI tools, 50–52
  BIM software, 48–49
  geographic information systems, 49
construction engineers, 13, **13**
co-ops, 19, 20

decision-making skills, 30–31
digital twin models, 50–51
drones, **47,** 49–50

earnings, 4, 38
earthquake-resistant structures, 14
Eason, Melinda, 14–15
educational requirements, 4
  ABET accreditation, 17–18
  classes
    for architects, 20
    in CAD, 23

before college, **22,** 22–24
in college, 18–19, 57–58
degrees, 17, 19
internships and co-ops, 19–21
for urban planners, 40
employers, 34–45
energy dissipation devices, 14
engineering services firms, 34
environmental projects
effects of climate change and, 46–47
examples of, 11–12
flood management and stormwater systems, 34
smart cities, 47, 51–52
sustainable, 15, 47, 48

Fern Hollow Bridge (Pittsburgh, Pennsylvania), 5–7
Frost, Jim, 12
Fuller, Jason, 6

generative AI, 50–51
geographic information systems, 49
geotechnical engineers, 14–15

Hall, James, 43–44

internships, 19–21
interpersonal skills, 29–30

James, Bill, 7
job advancement
expanding skills and, 41–42
PE license and, 40
typical steps in, 39
job descriptions
day in life of civil engineer, 56–57
day in life of structural engineer, 33–34
at a glance, 4
government policy, 41
locations, 35–36
teaching, 41
variety of tasks, 32
work schedules, 36–38
job outlook, 4, 45–46
job shadowing, 24
just-in-time design approach, 6

Kircher, Kelley, 15

licensing, 4, 21–22
Lim, Joan, 33–34
Love Building, Davidson–Davie Community College (Thomasville, NC), 20–21

McKinney, Mark, 10, 11
mentors/mentorships, 43–44
Mobed, Audrey, 20–21
Mondloch, Riley, 34, 49–50

networking, 40, 42–44

Occupational Outlook Handbook (Bureau of Labor Statistics), 34, 35, 38
O'Reilly, Finn, 18–19
organizational skills, 26–27, 40

Peña-Mora, Feniosky, 46
personal qualities and skills
  communication skills, 28–29, 40, 57
  decision-making skills, 30–31
  at a glance, 4
  interpersonal skills, 29–30
  problem-solving skills, 30, 57
  project management and organizational skills, 26–27, 40
  technical/hard skills, 25–26, 57
Prescott, Anna, 36
problem-solving skills, 30
Professional Engineer (PE) license
  advantages of having, 21
  career advancement and, 40
  as requirement, 4, 18, 21
  steps, 21
professional organizations, 40, 43
project engineers, 39
project management skills, 26–27
project management software, 27
public infrastructure projects, 35, **37,** 46
PVEDI Engineering, 56

Rashid, Shahab Naveed, 42–43
Revit, 26
Ridley, Rob, 16
Row, Shelley, 41–42

San Diego International Airport, New Terminal 1, 12
seismic dampers, 14
Siddiqi, Rizwan, 51
smart cities, 47, 51–52
Smith, Jamie, 32
social media, 43
spreadsheets, 27
Stephan, Jennifer, 18
structural engineers, 13, 33–34
surveyors/surveying
  described, 26, **28**
  drones and, **47,** 49–50
  requirements for career as, 23
Syed, Shaukat, 29

technical/hard skills, 25–26
testing designs using, 14
transportation engineers, 15
Tritico, Hans, 41
2025 Report Card for America's Infrastructure (ASCE), 46

urban planners/planning, 40

Venture, Jim, 56–57

Wyoming, 21

PICTURE CREDITS

Cover: eakkachai/Shutterstock

4: Shutterstock
13: peopleimages/iStock
22: monkeybusinessimages/iStock
28: NewSaetiew/Shutterstock
37: levkr/iStock
47: Frazao Studio Latino/iStock

ABOUT THE AUTHOR

Carla Mooney is the author of many books for young adults and children. She lives in Pittsburgh, Pennsylvania, with her husband and three children.